DOLPHINS

by Emma Bassier

Cody Koala

An Imprint of Pop!

popbooksonline.com

abdobooks.com
Published by Pop!, a division of ABDO, PO Box 398166, Minneapolis, Minnesota 55439. Copyright © 2020 by POP, LLC. International copyrights reserved in all countries. No part of this book may be reproduced in any form without written permission from the publisher. Pop!™ is a trademark and logo of POP, LLC.

Printed in the United States of America, North Mankato, Minnesota

052019
092019

THIS BOOK CONTAINS
RECYCLED MATERIALS

Cover Photo: Shutterstock Images
Interior Photos: Shutterstock Images, 1, 5 (bottom right), 7, 9, 15, 17, 19; iStockphoto, 5 (top), 5 (bottom left), 6, 20; Christopher Swann/Science Source, 10–11; Claus Lunau/Science Source, 12–13

Editor: Meg Gaertner
Series Designer: Jake Nordby

Library of Congress Control Number: 2018964497
Publisher's Cataloging-in-Publication Data
Names: Bassier, Emma, author.
Title: Dolphins / by Emma Bassier.
Description: Minneapolis, Minnesota : Pop!, 2020 | Series: Ocean animals | Includes online resources and index.
Identifiers: ISBN 9781532163388 (lib. bdg.) | ISBN 9781644940112 (pbk.) | ISBN 9781532164828 (ebook)
Subjects: LCSH: Dolphins--Juvenile literature. | Dolphins--Behavior--Juvenile literature. | Aquatic mammals--Juvenile literature. | Ocean animals--Juvenile literature.
Classification: DDC 599.53--dc23

Hello! My name is

Cody Koala

Pop open this book and you'll find QR codes like this one, loaded with information, so you can learn even more!

Scan this code* and others like it while you read, or visit the website below to make this book pop.

popbooksonline.com/dolphins

*Scanning QR codes requires a web-enabled smart device with a QR code reader app and a camera.

Table of Contents

Chapter 1

Water Mammal

Dolphins are smart water **mammals**. They live all over the world. Most dolphins live in the ocean. A few dolphins live in rivers.

Watch a video here!

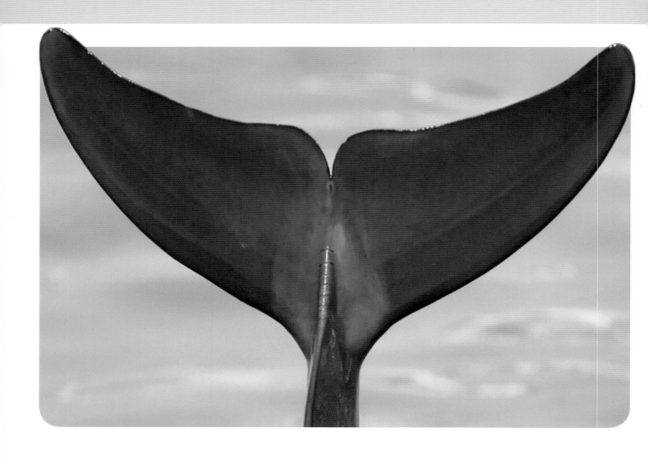

Dolphins have two **flukes** that make a strong tail. Flukes help dolphins swim fast.

Dolphins also have **flippers**.
The two flippers help them
change direction.

Dolphins live in water.
But they breathe air.
Dolphins swim to the water's
surface. They breathe
through a **blowhole** on top
of their heads.

A dolphin's blowhole closes after each breath. This helps keep water out.

Chapter 2

Fish Food

Dolphins eat fish and other animals. Many dolphins hunt in groups called pods. A pod swims around a group of fish. The dolphins take turns hunting and eating the fish.

Learn more here!

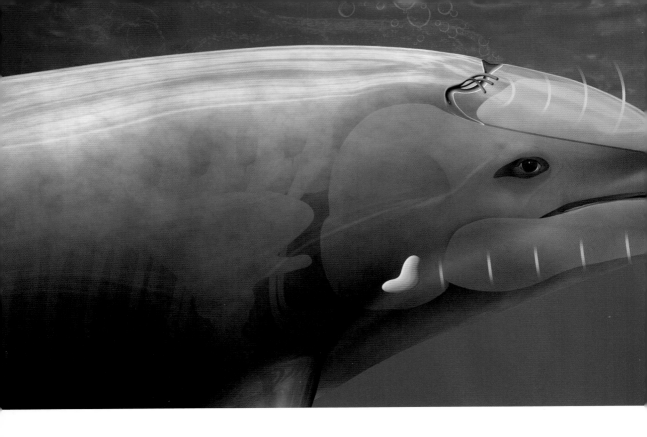

Dolphins also hunt using **echolocation**. The dolphin makes a sound. The sound bounces off other objects.

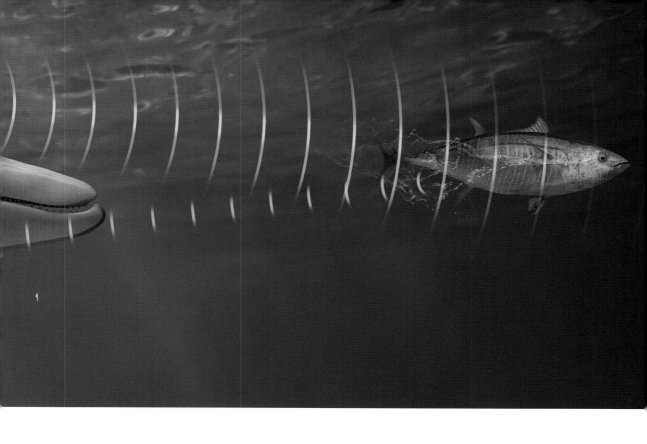

The dolphin listens to the echoes. It can tell how far away the objects are.

Chapter 3

Social Swimmers

Dolphins are smart, playful animals. They **whistle** to talk to one another. They work together to stay safe from sharks.

Learn more here!

Dolphins also work together to help sick or hurt dolphins. They bring the hurt dolphin to the water's surface. That way, the hurt dolphin is able to breathe.

A pod can include between five dolphins and several hundred dolphins.

Chapter 4

A Dolphin's Life

Mother dolphins have one baby at a time. The baby dolphin is called a calf. The calf stays with its mother for three to eight years.

Complete an activity here!

The calf learns to hunt and stay safe. Then it joins a new pod. Most dolphins live to be 30 years old. But some kinds can live to be 50 years old.

Making Connections

Text-to-Self

Have you ever seen a dolphin close up? Would you want to? Why or why not?

Text-to-Text

Have you read about another animal that uses echolocation to hunt? How is that animal similar to a dolphin? How is it different?

Text-to-World

Dolphins live in pods. What other animals live in groups? How does living in a group help some animals?

Glossary

blowhole – a hole on the top of a dolphin's head that is used for breathing.

echolocation – the process of using sound to locate objects.

flipper – one of two fins on the sides of a dolphin that help it swim.

fluke – one of two sections on a dolphin's tail.

mammal – an animal that breathes air and whose babies drink milk produced by their mothers.

whistle – to make a clear, high-pitched sound.

Index

Online Resources

popbooksonline.com

Thanks for reading this Cody Koala book!

Scan this code* and others like it in this book, or visit the website below to make this book pop!

popbooksonline.com/dolphins

*Scanning QR codes requires a web-enabled smart device with a QR code reader app and a camera.